Prehistoric marine
life in Australia's inland sea

Platypterygius australis
Artist: Frank Knight

Museums Victoria
Nature Series

Prehistoric marine
life in Australia's inland sea

Danielle Clode

MUSEUMSVICTORIA
PUBLISHING

Prehistoric marine
life in Australia's inland sea

Published by
Museums Victoria Publishing

First printed 2015, reprinted 2024

Museums Victoria Publishing
GPO Box 666
Melbourne VIC 3001 Australia
+61 3 8341 7536
museumsvictoria.com.au/books

PRINTED IN
China through Asia Pacific Offset

SERIES DESIGN BY
Stephen Horsley, Propellant

DESIGNED & TYPESET BY
Elizabeth Dias, studioether

National Library of Australia
Cataloguing-in-Publication entry

CREATOR: Clode, Danielle.
TITLE: Prehistoric marine life in
Australia's inland sea / Danielle Clode.

ISBN: 9781921833168 (paperback)

NOTES: Includes bibliographical
references.

TARGET AUDIENCE:
For secondary school age.

SUBJECTS:
Marine animals, Fossil--Eromanga
Basin--Pictorial works. Animals,
Fossil--Eromanga Basin--Pictorial
works. Eromanga Basin.

DEWEY NUMBER:
560.450994

COVER IMAGE:
Polycotylid plesiosaur giving birth. Drawing by
Stephanie Abramowicz, Dinosaur Institute,
Los Angeles County Museum of Natural History.

Sphenopteris warragulensis specimen from the Cretaceous period found in Koonwarra, Victoria.
Photographer: Benjamin Healley, Museums Victoria

CONTENTS

Artist's interpretation of a pliosaur catching a pterosaur.
Artist: Tor Sponga, Bergens Tidende

The inland sea

A view from Cape York

Rocky outcrops in Australia's north once looked out over an inland sea instead of dry land.
Photographer: Kyle Taylor

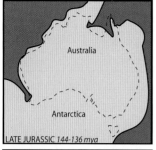

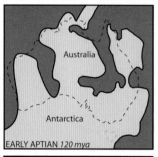

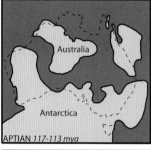

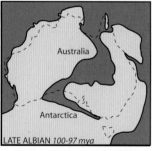

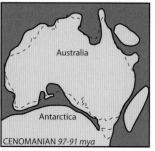

From the top of the granite escarpment an endless plain stretches to the horizon. Sparse eucalypt woodlands shimmer in the heat. The uniform olive green of the plains is broken only by the occasional rocky outcrop and the darkened lines of creeks and rivers meandering across the wide flat interior of north-western Cape York Peninsula.

From this vantage point at the top of Australia, there are no signs of modern human life. There are no towns or cities: no cars, no roads, no fences, not even the telltale vapour trail of a distant jet. The vast space is filled with silence, broken only by the echo of a crow and a whisper of wind in the rocks. This is a primeval landscape, one that seems to have remained unchanged for aeons.

The rock on which we sit is part of the ancient igneous spine that runs from Cape York down the eastern edge of the Australian continent, stubbornly resisting the onslaught of wind and water over millions of years. Rich tropical rainforests cloak the range along the coast to the east. To the west and south, lies the dry inland, stretching low and flat into the desert country of Australia's red heart.

The changing Earth

The apparently ancient and unchanging appearance of the landscape is deceptive. In the last days of the Jurassic, 150 million years ago, the same viewpoint would have revealed a forest, not of dry-land eucalypts, but lush tropical conifers, cycads and ferns. In the distance, we might have heard not the cries of crows but the bellowing of sauropods plodding along the open floodplains of a riverbed.

For millions of years life had flourished in a warm and wet Jurassic world. But the climate was changing. Temperatures were falling and the great Gondwanan land mass was wrenching

As sea levels changed from the Late Jurassic through to the middle of the Cretaceous, Australia's inland sea expanded and contracted. (Adapted from Frakes 1987)

itself apart. The tectonic plates of the Earth's crust shifted and realigned, thrusting mountain ranges upwards, creating the Rockies in North America, the Andes in South America and the Alps in Europe. In Australia, sections of the western plateau subsided, creating meandering river systems and large shallow lakes.

The vast network of rivers and lakes that had stretched across Australia during the Jurassic began to crust with ice in winter as the temperatures fell in the Cretaceous. Glaciers formed on the mountains, depositing ice-scarred boulders and dropstones into the ever-expanding shallow lakes. As the sea levels rose, the lakes turned into oceans and the entire inland of Australia—what is today

Early Cretaceous plant community.
Artist: Karen Carr

3

desert—was flooded by the Eromanga Sea. In the middle of the Cretaceous—the Aptian age of 120 million years ago—we would not have been looking out over the vast sweeping plains of a hot, flat continent, but perched on the edge of one island among many, surrounding a cold, shallow sea. Australia was not so much a continent as a scattered archipelago surrounding a vast inland ocean.

Warm and dry or cold and wet?

Fossil of *Ginkgoites australis*.
Photographer: Benjamin Healley, Museums Victoria

From the Aptian to the Albian, from 125 to 100.5 million years ago, Australia's red centre was a cool blue. Sea levels rose and fell, the climate warmed and cooled, animals appeared and disappeared. Along the northern shores of this inland sea, conditions were warm and dry. Cycads flourished, along with palm-like plants, ferns and conifers. On the edge of the calmer lagoons, the forest grew right down to the water's edge, dominated—as were Cretaceous forests worldwide—by towering Araucaria pines. Today, the modern descendants of these forests, such as the New Zealand Kauri, Norfolk Island Pine and Bunya Pine, are restricted to the continents of Australia and South America, and the islands in between.

Among these giants grew ancient gingko trees, almost identical to the ones still growing slowly in Asian forests today, along with a diverse array of cycads. Less familiar would have been the strange-looking horsetail plants. The undergrowth sheltered an abundance of royal ferns, relatives of the modern king ferns that still flourish in our east coast forests. Insect life was abundant and diverse, particularly beetles and aquatic nymphs.

Bunya Pine *Araucaria bidwillii*
Photographer: Danielle Clode

PREHISTORIC MARINE LIFE IN AUSTRALIA'S INLAND SEA

Eon	Era		Period	Epoch
Phanerozoic	Cainozoic		Quaternary	Holocene
				Pleistocene
			Neogene	Pliocene
				Miocene
			Palaeogene	Oligocene
				Eocene
				Paleocene
	Mesozoic		Cretaceous	Late — ALBIAN
				Early — APTIAN
			Jurassic	
			Triassic	
	Palaeozoic	Late	Permian	
			Carboniferous	
			Devonian	
		Early	Silurian	
			Ordovician	
			Cambrian	
Pre Cambrian	Proterozoic			
	Archaean			

Geological timescale

Invertebrate life flourished around the freshwater lakes and streams—dragonflies, caddis flies, mayflies and scorpion-flies emerged from their aquatic infancy and took to the air. Beetles, fleas, millipedes and harvestman spiders populated the overhanging vegetation, while earthworms, yabbies and bivalves buried into the mud. Colonies of bryozoan moss animals, or lace corals, filtered their food from the cold clear waters.

On the southern coasts, wet conditions prevailed and the water was near freezing. Ice crusted the southern coastlines of the Eromanga Sea. The conifer and cycad forests struggled with seasonal changes in rainfall, perhaps opening the way for the emergence of new types of plant: ones with flowers.

At the forest edge, along the shore of the Eromanga Sea, things might not immediately have felt so different from today. The creatures flying overhead might be pterosaurs rather than pelicans, but the beach would still have been strewn with the oceanic flotsam recognisable today: cockles and crabs, seaweeds and sponges.

Under the water's surface, the sea floor may well have looked quite familiar too. Delicate and beautiful sea lilies or crinoids, *Isocrinus australis*, decorated the ocean as they have for millennia, their colourful plumed arms sifting out particles of food wafting by on the current. The brain coral *Mckenziephyllia accordensis* formed large rounded domes with honeycombed or grooved surfaces, looking rather like fluorescent brains, and the widespread and abundant sponge *Purisiphonia clarkei* poked

Fossil and reconstruction of a crinoid. Photographer: Danielle Clode. Source: South Australian Museum

long finger-like structures from the sea floor, coloured yellow and orange by the bacteria and dinoflagellates it filtered from the water.

A range of crustaceans scuttled across the sea bed, including the clawed lobster *Hoploparia mesembria*, which was rather like the large modern American and European clawed lobsters. Squat little crabs, like *Torynomma quadrata*, fed on detritus on the shoreline and sea floor. *Homolopsis etheridgei* was a member of the carrier or porter crab family, characterised today by their habit of camouflaging themselves with plants, coral and even sea urchins carried over their backs with specially formed rear legs. Schools of shrimp (*Mecorchirus*) may well have drifted through the Albian waters in similar abundance to today.

But swimming in the waters of the inland sea, we might have encountered very unfamiliar creatures indeed. We might have drifted with ammonites, looking bizarrely as if a squid had taken up residence, hermit-like, in a floating shell. Armoured fish schooled in the deeper waters, as if on guard against some hidden threat. Shadows drifting in the distance might remind us of whales. But these monsters were no peaceable plankton-feeders. Long sinuous necks indicate a creature more akin to a Loch Ness monster—plesiosaurs hunting for belemnites and ammonites. A flurry of darting, desperate fish are driven forwards by a school of ichthyosaurs, their dolphin-like shape belying their much larger size and reptilian origins. And in turn the ichthyosaurs are hunted by even larger beasts, predatory reptiles about the size of a school bus—the gigantic pliosaur *Kronosaurus*.

It is almost impossible to imagine this cold alien ocean which once submerged the red deserts of modern Australia. As we look out over the flat plains of inland Australia today, there is little sign of the dramatic changes wrought to a maritime kingdom 100 million years ago. The keys to this past world lie hidden, deep in the rocks, many still waiting to be uncovered.

Fossil sites

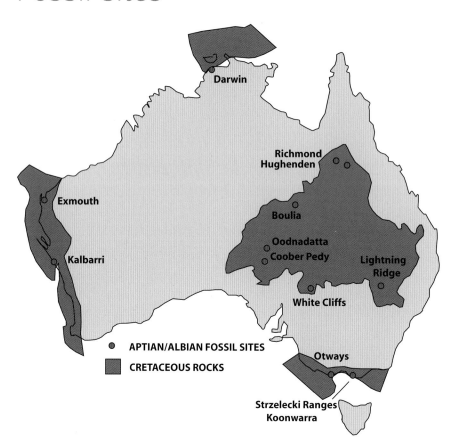

Cretaceous rock formations with major Aptian/Albian fossil sites

Aquatic fossils from the middle of the Cretaceous in Australia are found not only on the former floor of the inland sea, the Eromanga Basin, but also from marine and freshwater deposits on the perimeter of the continent, on the coast of Western Australia and the Northern Territory and from river deposits in Victoria.

In the Eromanga Basin, there are major fossil sites in the South Australian opal-mining districts of Coober Pedy and Andamooka as well as nearby Curdimurka and Oodnadatta. White Cliffs, in New South Wales, is also an opal-mining town, and a source of fossils from the Aptian–Albian. In the north of the basin, the western Queensland town of Winton is famous for its dinosaur fossils, but nearby Hughenden, Richmond, Julia Creek, Boulia and Prairie yield an even richer array of marine fossils from the same period.

Koonwarra fossil. Photographer: Professor James Warren

The Surat Basin adjoins the Eromanga Basin to the east. Together these formations create the Great Artesian Basin, the largest and deepest freshwater basin in the world covering an area of over 1.5 million square kilometres. The Surat Basin covers much of inland New South Wales and contains major fossil sites near Roma, Surat and Lightning Ridge.

Marine deposits on what are now coastal regions of Australia have also provided significant fossils for the same period as the inland sea. In Victoria, the Otway Ranges and Strzelecki Ranges have produced terrestrial, freshwater and marine fossils from the middle of the Cretaceous. In Gippsland, the Koonwarra fossil beds, uncovered during road works on the South Gippsland Highway, provide one of the most significant collections of freshwater fish, plants and insects from the mid-Cretaceous.

In the Northern Territory, the Money Shoals Basin yields abundant marine fossils—reptiles, sharks and fish—around the Darwin suburbs of Casuarina Beach, Fannie Bay and Nightcliff.

Many of the Western Australian fossil beds exposed by erosion on the coast contain fossils from a wide array of ages. Examples of the earliest forms of multicellular life—stromatolites—can be found near the track ways of some of the earliest organisms to walk on land in areas such as in the Murchison River region. Further south, megafaunal deposits overlap with human occupation at Devil's Lair, Cape Leeuwin. Other coastal sites, stretching from Perth to Exmouth, are significant for their records of Cretaceous marine life. The Giralia Ranges, in particular, document in stark relief the impact of the Cretaceous–Tertiary extinction event, which marks the end of the age of the dinosaurs.

Glendonites, dropstones and opals

The conditions for fossilisation are rare indeed, but the Eromanga Sea is home to even more unusual geological formations than just fossil-bearing deposits. These crystals, stones and gems are unique to the peculiar climatic and geological conditions that lead to the creation and preservation of the fossils from this period in Australia's past.

Glendonites are the remains of star-shaped crystals that only form in cold alkaline waters. The original crystals are made from a waterlogged form of calcium carbonate called ikaite, which is only stable in near freezing water and 'dehydrates' to form calcite crystals at warmer temperatures. Although the original crystals are rare and difficult to preserve, their characteristically spiky form is often replaced by other more durable substances and the resulting rock crystals are often named after the location in which they were originally found. Glendonites were first found in Glendon in New South Wales. But in other places they are known as hedgehogs and rose rocks. At White Cliffs, the calcite is often replaced by opal, resulting in 'opal pineapples'.

Glendonite fossils are commonly found in the South Australian Bulldog Shale of the Eromanga Sea, suggesting that water temperatures here were between −2 and 7° Celsius. The glendonites form on or just below the surface of the sea floor, in the rich organic mud that layered the ocean floor in the Aptian. These glendonites are often found where there is also evidence of ice-rafts or glaciers, such as dropstones.

The characteristically smooth surface of dropstones reveals a history of being rubbed against one another by the tumbling, grinding action of water. But unlike pebbles in a riverbed or along a coastline, dropstones are often large (up to 2 metres across) and found in the middle of oceans, far from any possible source of friction. The way they reach their destination is remarkable, for dropstones are created in rivers of ice, which pick up rocks as they

Display of glendonites and dropstones.
Photographer: Danielle Clode.
Source: South Australian Museum

crawl over the land. When these glaciers reach the sea, they break into icebergs which float away, slowly melting and releasing stones and rocks mid ocean. Other smaller rocks, found in isolation, may be mistaken for dropstones. These include stomach stones worn smooth by marine reptiles, either to help them digest their food or to provide ballast.

One of the most striking geological features of Eromanga Sea fossils, and the rocks containing them, are the opals. The beautiful and diverse colours of these gems are created when light is diffracted by the silica spheres that make up natural opal. Opals, particularly of gemstone quality, are rare and the vast bulk of them come from Australia. In fact, more than 90 per cent of the world's mined opal comes from the Great Artesian Basin of Australia— the area once filled by the Eromanga Sea.

Unusual conditions are required to create opal. For many years these were poorly understood and hotly debated. But recent research suggests that opal formation requires a large volume of sediment rich in iron and organic material to be washed into a low oxygen and low carbonate environment, for example by a vast network of rivers pouring into a shallow inland sea. When the sea recedes, the sediments are exposed and then 'rust'. This oxidative weathering also gives red desert soils their characteristic colour.

Display of opalised bivalves from Eromanga seabed, Coober Pedy. Photographer: Danielle Clode. Source: South Australian Museum

Finally, the sediments need to be covered by more sediment to protect them from further erosion and allow what has become silica gel to be transformed into quartz. Precisely these conditions appear to have occurred with the inundation, then retreat, of the Eromanga Sea, leading to the opalisation of many of the fossils of the creatures that once lived there. The only other place that seems to have undergone similar conditions for widespread opalisation is on Mars. Who knows if any fossils might be preserved in precious gemstone there?

Opalised fossil of a gastropod shell from Coober Pedy, South Australia. Photographer: Frank Coffa. Museums Victoria

Inhabitants of
the inland sea

Giants of the sea

Aristonectes parvidens feeding. Artist: Brian Engh

Plesiosaurs

KINGDOM: Animalia

PHYLUM: Chordata

CLASS: Reptilia

SUPERORDER: Sauropterygia

ORDER: Plesiosauria — near lizard

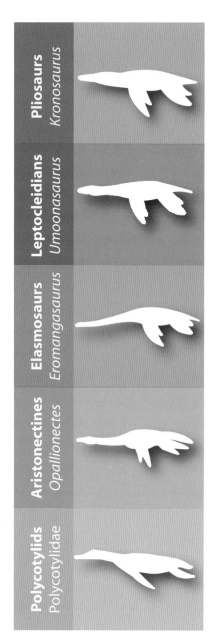

Pliosaurs
Kronosaurus

Leptocleidians
Umoonasaurus

Elasmosaurs
Eromangasaurus

Aristonectines
Opallionectes

Polycotylids
Polycotylidae

Plesiosaurs were one of the dominant predators of Mesozoic oceans around the world. These gigantic marine reptiles had flippers like seals but grew up to 13 metres long—about the size of a modern sperm whale. Plesiosaurs are best known for their resemblance to the mythical Loch Ness monster, but they were actually quite diverse in shape, ranging from elegant long-necked elasmosaurs to the massive, big-headed pliosaurs.

When first discovered, plesiosaur bones and fragments were thought to be those of fish, crocodiles or dolphins. The first almost complete skeleton was found at Lyme Regis on the coast of Dorset in 1823 by Mary Anning, a fossil collector who discovered the first ichthyosaur, with her brother Joseph, at the age of twelve. Mary Anning's discovery led to the plesiosaurs being recognised as belonging to their own group, Plesiosaurus. Early depictions often showed plesiosaurs sunning themselves on the shore and curving their long necks, swan-like, in the air. In fact, plesiosaur flippers seem to be much better adapted for an entirely aquatic life and were unlikely to have been strong enough to lift large bodies onto land. Similarly, their long necks had limited flexibility and could not have been supported out of water.

In Australia, plesiosaur remains are often found in the southern regions of the inland sea, including freshwater areas of the Otway Basin in Victoria. Here, temperatures were near freezing and ice may have encrusted the coast and rivers. Recently, scientists have found evidence that many of these large aquatic reptiles were able to maintain higher body temperatures than the surrounding water, which may explain their abundance and success in the cold southern oceans.

Front view of the Addyman plesiosaur skull reconstruction.
Photographer: Danielle Clode.
Source: South Australian Museum

Pliosaurs

SUPERORDER: Pliosauroidea

— more lizard (than the ichthyosaur)

SPECIES	ETYMOLOGY	LENGTH
Kronosaurus queenslandicus	King of the titans from Queensland	10.5 m

Kronosaurus queenslandicus was the largest marine reptile to cruise the waters of the Eromanga Sea, reaching lengths of around 10 metres and weighing up to 11 tonnes. The skull of this huge predator was over 2 metres long and equipped with conical interlocking teeth 15 centimetres long for crushing the shells of ammonites, turtles and armoured fish. The flattened skull and upward-facing eyes of *Kronosaurus* may have helped it to ambush prey from below. *Kronosaurus* is known to have eaten other plesiosaurs and sharks, including a large elasmosaur that may have been 8 metres long. In general though, *Kronosaurus* probably consumed prey smaller than itself, including shellfish.

Kronosaurus fossils are not common, but are found from both the Albian and Aptian. The first *Kronosaurus* fossil—a section of jaw—was found near Hughenden, central Queensland, in 1899. An almost complete skeleton was located nearby in 1931 on Army Downs. Station owner Ralph Thomas told a visiting American team of palaeontologists about the bones and they eventually

Kronosaurus queenslandicus. Artist: Josh Lee, Adelaide.
From *Dinosaurs in Australia, Mesozoic life from the Southern continent*, by Benjamin P. Kear & Robert J. Hamilton-Bruce

returned to the Harvard Natural History Museum with 5 tons of fossil material for the specimen. It took twenty years to prepare and mount the specimen for display, with funding coming from a wealthy American family whose ancestors had once reported the existence of a sea monster off the eastern American coast. Ralph Thomas eventually went to America at the age of ninety-three to see his discovery. More complete skeletons have since been found in Australia.

Leptocleidians

FAMILY: Leptocleididae — Slender clavicle

SPECIES	ETYMOLOGY	LENGTH
Leptocleidus clemai	Slender clavicle, after John M. Clema	3 m
Umoonasaurus demoscyllus	Umoona (Coober Pedy region) people's sea monster	2.5 m

Leptocleidians were a widespread family of small plesiosaurs. Their remains are often found close to shore or in estuaries, and some leptocleidians seem to have lived even further inshore, where their fragmentary remains have been found in freshwater river sediments. With their relatively small size and four large flippers, leptocleidians may have occupied an environmental niche similar to modern seals.

The crests on the skulls of *Umoonasaurus* may have supported a horny covering used in displays like marine iguanas. *Umoonasaurus demoscyllus*.
Artist: Josh Lee, Adelaide. From *Dinosaurs in Australia, Mesozoic life from the Southern continent*, by Benjamin P. Kear & Robert J. Hamilton-Bruce

The Australian leptocleidians are best represented by a beautifully opalised young adult *Umoonasaurus* known as 'Eric', which is now on display at the Australian Museum. In life, leptocleidians carried their own ornamentation. *Umoonasaurus* appears to have had a crest on its head, which may have been used for display.

Elasmosaurs

FAMILY: Elasmosauridae

— ribbon lizard

SPECIES	ETYMOLOGY	LENGTH
Eromangasaurus australis	Southern Eromanga lizard	9–10 m

Elasmosaurs were once assumed to be cosmopolitan and found around the world. Closer inspection, however, has revealed that the Australian elasmosaur species were quite distinct from their overseas cousins. Much of the elasmosaur fossil record is too fragmentary to assign to a species, with the exception of *Eromangasaurus australis*. This particular species has been identified on the basis of a skull that is notable because it was crushed by a large predator, presumably *Kronosaurus*.

Eromangasaurus australis
Artist: Josh Lee, Adelaide. From *Dinosaurs in Australia, Mesozoic life from the Southern continent*, by Benjamin P. Kear & Robert J. Hamilton-Bruce

Elasmosaurus plesiosaur. Artist: Frank Denota. Copyright Dorling Kindersley

The long neck of the elasmosaurs has given rise to much debate over the way in which they fed. Their neck may have allowed them to feed on the sea floor, or to get closer to fish without being noticed. Some have suggested that they might be able to strike at fish, like snakes or turtles do, or even swipe at them sideways like saw sharks. The strange protruding conical teeth of the elasmosaurs suggest that they specialised in fast-moving fish and squid, but two Australian specimens have been found with stomach contents dominated by bottom-dwelling shellfish and crustaceans, which they probably ground up with the aid of small pebbles, known as gastroliths, stored in their stomachs.

Aristonectine plesiosaurs

FAMILY: Aristonectidae — best swimmer

SPECIES	ETYMOLOGY	LENGTH
Opallionectes andamookaensis	Opal swimmer of Andamooka	5 m

The Aristonectine plesiosaurs may have been a kind of elasmosaur and were very common in the southern hemisphere. They occurred in the Antarctic, New Zealand and Patagonia as well as the southern areas of the Eromanga Sea, suggesting that they may have been particularly adapted for low temperatures.

Their fossils are also found in Europe and North America. Aristonectes from Antarctica have fine, outward facing interlocking teeth, which may have been used to strain krill from the water. The Australian aristonectid *Opallionectes andamookaensis* had small oval-shaped needle-like teeth, which may also have been adapted for filter feeding.

Head of *Aristonectes parvidens*
Artist: Brian Engh

Aristonectes parvidens with ammonites. Artist: Dmitry Bogdanov

Polycotylid plesiosaurs

FAMILY: Polycotylidae — much cupped vertebrae

SPECIES	ETYMOLOGY	LENGTH
Polycotylidae sp.	With much-cupped vertebrae	5 m

Fossil remains of these long-nosed, short-necked plesiosaurs are rare from the Eromanga Sea. But the fact that they were present at all is even more remarkable. Most polycotylids are found in the northern hemisphere, in the Late, rather than Early Cretaceous. These as yet unnamed Australian forms, represented by an opalised partial skeleton from New South Wales and an almost complete skeleton from Richmond in Queensland, may be the oldest known members of their family in the world.

Polycotylid plesiosaurs gave birth to large, live young, rather than laying eggs.
Drawing by Stephanie Abramowicz, Dinosaur Institute, Los Angeles County Museum of Natural History

In 1987, the remains of a large adult polycotylid plesiosaur were found in Kansas, surrounding those of a smaller immature individual, suggesting a pregnant mother. This unborn plesiosaur was surprisingly large—1.5 metres long—compared to the mother's length of 4.7 metres. Giving birth to live young, rather than laying eggs on land as many other marine reptiles do, allowed the plesiosaurs to adapt to a more exclusively aquatic life. This trait may also have made them more social. While modern marine reptiles (like crocodiles and turtles) have many small young, plesiosaurs seem to have developed a family life more like that of modern whales, giving birth to a single large young which may have been cared for over a longer period, perhaps in a social group. As a result, it is possible that plesiosaurs of the Eromanga Sea may have lived in family groups, caring for and protecting their young and perhaps teaching them how to capture prey.

Ichthyosaurs

KINGDOM: Animalia
PHYLUM: Chordata
CLASS: Reptilia
CLADE: Eoichthyosauria
ORDER: Ichthyosauria — fish lizard

SPECIES	ETYMOLOGY	LENGTH
Platypterygius australis	Australian broad-fin	5 m

Ichthyosaurs were large dolphin-like aquatic reptiles with species ranging from 1 to 16 metres in length. Early on, ichthyosaurs were known as 'sea dragons' and were often depicted sunbathing on rocks, with long straight tails, despite a 'kink' often seen in fossils. It wasn't until some beautifully preserved fossils were found in Germany, preserving the outline of the ichthyosaurs, that scientists realised that the 'kink' in the tail was actually part of a lower tail fluke. The upper part of the tail and dorsal fin have no bones and so are rarely preserved in fossils. The ichthyosaurs were clearly even more fish-like than scientists first thought.

In 1846, English surgeon and fossil collector, Joseph Chaning Pearce, first described a fossil of an ichthyosaur apparently giving birth to young. Hundreds

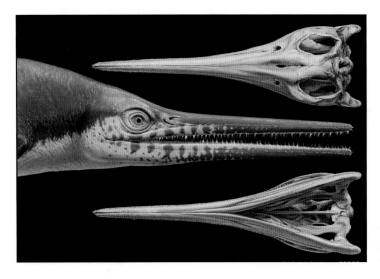

Reconstruction of the ichthyosaur
Platypterygius australis showing
the developmental drawings of
bone and cartilage muscle.
Artist: Peter Trusler

Platypterygius australis
Artist: Josh Lee, Adelaide. From *Dinosaurs in Australia, Mesozoic life
from the Southern continent*, by Benjamin P. Kear & Robert J. Hamilton-Bruce

of fossilised Jurassic ichthyosaurs have since been found in Germany, in varying
stages of pregnancy. These pregnant females allow us to tell the fossils of males
from females, revealing that the sexes were similar in size.

Ichthyosaurs were found across the world, first appearing 245 million years
ago, and disappearing 25 million years before the end of the Cretaceous, when
the dinosaurs became extinct. The abundant ichthyosaurs of Australia's inland
sea appear to have been amongst the last of their kind, and more common in the
Albian than they had been in the Aptian.

The Australian ichthyosaur *Platypterygius australis* was a fairly large species,
growing up to 5 metres long. Its articulated fins were long and broad, built for

Ichthyosaurs hunting baby sea turtles. Artist: Josh Lee, Adelaide. From *Dinosaurs in Australia, Mesozoic life from the Southern continent*, by Benjamin P. Kear & Robert J. Hamilton-Bruce

27

fast cruising rather than for sprinting, and used primarily for steering. Like the plesiosaurs, the ichthyosaurs seem to have been warm-blooded, giving them an important advantage in the cold Eromanga Sea.

The sensory capacity of the ichthyosaurs is quite well studied. The Australian ichthyosaur had a long slender skull with characteristically large eyes. In fact, a European ichthyosaur has the largest eyes recorded for any animal, up to 26 centimetres across. Their vision was probably excellent, given the size of the optic lobes in their brains, and it is likely that they used vision rather than echolocation to hunt their prey. They also seem to have had an excellent sense of smell and may have had electroreception, the ability to sense natural electrical fields. Large eyes, enhanced smell and electroreception are all adaptations to aquatic hunting under low light conditions, although they may also have other advantages. Electroreception in sharks, for example, might also be used for navigation.

Platypterygius australis had jaws set with numerous large conical teeth that meshed together when they closed. These teeth, which snap shut but don't chew, suggest they ate their food whole and probably hunted relatively small, fast-moving prey. This is confirmed by fossilised stomach contents, which include fast-swimming, herring-like fish (*Pholidophorus*), squid-like belemnites, young turtles and even birds.

Platypterygius australis skull specimen. Photographer: Jon Augier, Museums Victoria

Animals of land and water

Turtles eating jellyfish. Artist: Maree Maxwell, Kronosaurus Korner

Turtles

KINGDOM: Animalia

PHYLUM: Chordata

CLASS: Reptilia

CLADE: Testudinata

ORDER: Testudines — shelled

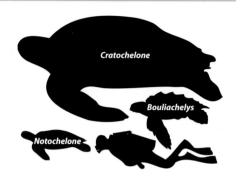

SPECIES	ETYMOLOGY	LENGTH
Marine turtles		
Notochelone costata	Ribbed southern turtle	Less than 1 m
Bouliachelys suteri	Boulia turtle, after local fossil collectors Richard and John Suter	More than 1 m
Cratochelone berneyi	Powerful turtle after Frederick L Berney, naturalist and owner of Barcarolle Station QLD	3–4 m
Freshwater turtles		
Chelycarapookus arcuatus	'Bow-like' turtle from Carapook parish	20 cm
Otwayemys cunicularis	Otways turtle found by mining	22 cm
Spoochelys ormondea	Turtle from Spook's Field named after collector Ormond Molyneux	30 cm

Bouliachelys suteri
Artist: Josh Lee, Adelaide. From *Dinosaurs in Australia, Mesozoic life from the Southern continent*, by Benjamin P. Kear & Robert J. Hamilton-Bruce

Turtles, like crocodiles, are a branch of the archosaur group, which includes dinosaurs, pterosaurs and birds. Modern turtles proliferated in the oceans during the Mesozoic period with a great radiation of species in the Jurassic and into the Cretaceous. Their evolutionary history is complex and poorly understood, but the abundance of such heavily armoured marine animals in the Mesozoic oceans tells us much about the evolutionary pressure that large marine predators, like the plesiosaurs, may have exerted on a wide range of prey species. In today's oceans, turtle shells

Modern leatherback turtle, *Dermochelys coriacea*.
Photographer: Brad Maryan

remain somewhat over-engineered, but they are the survivors of an ancient war with long extinct predators, and still wear the armour that gave them their decisive advantage.

Many turtle families became extinct along with the dinosaurs and many marine lineages at the end of the Cretaceous. Other families, like the horned tortoises, survived until relatively recently, only to become extinct in the Pleistocene extinctions when much of our megafauna disappeared. Some of our modern turtle families, however, had relatives who swam in the inland waters of the Eromanga Sea, much as they do along our coastline today.

Notochelone costata is the most commonly found fossil turtle from the Eromanga Sea, and may even have been more common than their larger and more charismatic relatives, the plesiosaurs and ichthyosaurs. *Notochelone costata* was a uniquely Australian turtle but has no living descendants today. They were about the size of modern turtles—less than a metre in length—and may have been similar to a modern hawksbill turtle. A slightly larger turtle, *Bouliachelys suteri*, probably grew to over a metre long.

Modern turtles have very diverse diets. Green turtles eat sea grass and algae, but other species are specialist feeders on jellyfish, crustaceans or shellfish. The ancient turtles of the Eromanga Sea may have had similarly diverse diets. *Bouliachelys* was once assumed to have eaten hard-shelled prey like ammonites—judging from their crushing plates and serrated hooked beak. A recent study of gut contents of fossil Australian sea turtles, however, shows that they actually fed on the abundant bivalves that covered the shallow sea floor in the Albian.

The largest of all the turtles at the time was *Cratochelone berneyi* of which just a single fossil has been found in the inland sea, suggesting that it may have been a rare visitor from the open oceans. This turtle grew up to 3 or 4 metres long. Leatherback turtles, which are the closest modern equivalent, are the largest modern marine turtle at an average of 1.5 metres for adults; however, the largest ever recorded was 3 metres long and weighed nearly 1 tonne. A giant Mesozoic turtle, *Archelon ischyros*, from the American Western Interior Seaway, typically grew to a similar size, the largest individual being nearly 5 metres long, possibly weighing almost 2 tonnes. The bones of *Cratochelone berneyi* also had a complex network of blood vessels, like modern leatherback turtles, suggesting that it may have been able to grow to a large size very rapidly.

Not all turtles lived in the oceans though. Turtles could also live in freshwater and on land. *Chelycarapookus arcuatus* was a small turtle, just 20 centimetres long, commonly found in the Victorian rivers of the Aptian period. *Otwayemys cunicularis*, found at Dinosaur Cove in Victoria, was slightly larger and may have been related to the giant horned tortoises along with *Spoochelys ormondea*, which seems to have been adapted to a more terrestrial lifestyle, with hoof-like claws and a broad snout.

Crocodiles

KINGDOM: Animalia
PHYLUM: Chordata
CLASS: Reptilia
SUPERORDER: Crocodylomorpha
ORDER: Crocodilia — stone worm

SPECIES	ETYMOLOGY	LENGTH
Isisfordia duncani	From Isisford QLD, after discoverer Ian Duncan	1.1 m

Crocodiles, like turtles, dinosaurs, pterosaurs and birds, are a branch of the archosaur group. Like turtles, the crocodiles emerged and diversified in the Mesozoic era, taking to an aquatic life and sometimes growing to enormous sizes. Some of the largest crocodiles ever known date from the middle of the

Cretaceous. *Sarcosuchus imperator*, for example, was the size of a bus—12 metres long and weighing 8 tonnes: a generalist predator inhabiting the river systems of central Africa. In America, the Late Cretaceous *Deinosuchus* grew to similar sizes.

Although Australia seems to have missed out on these crocodilian giants, it was home to the very earliest modern crocodile. The relatively tiny *Isisfordia duncani*, found near Winton in Queensland, may only have grown to a little over a metre long, but its presence suggests that crocodiles may well have first

Isisfordia duncani. Artist: Matt Herne

evolved in Gondwana, rather than North America. *Isisfordia* shows some of the adaptations that allowed modern crocodiles to become large effective predators both in water and on land.

Fragments of other crocodiles have also been found, particularly in the warmer northern areas of the Eromanga Sea. Certainly more than one type of crocodile lived in or around the inland sea. One fossil fragment from Lightning Ridge has a broad snout with slicing teeth typically used to shear meat, while another fragment has the typically conical teeth of a fish-eater.

Amphibians

KINGDOM: Animalia

PHYLUM: Chordata

SUPERCLASS Tetrapoda

CLASS: Amphibia — both kinds of life, terrestrial and aquatic

SPECIES	ETYMOLOGY	LENGTH
Koolasuchus cleelandi	'Crocodile' named after collector and preparators Lesley Kool and Mike Cleeland	4–5 m

Koolasuchus cleelandi jawbone. Photographer: Andrew Curtis. Museums Victoria

Koolasuchus cleelandi. Artist: Peter Trusler

Today's amphibians tend to be small and rather innocuous aquatic creatures—represented by the frogs and toads, salamanders and the little-known caecilians. But the ancestors of modern amphibians—more precisely known as temnospondyls—were highly diverse, abundant and sometimes very large indeed. As the first vertebrates to begin colonising the land, long before the rise of the reptiles, amphibians evolved a range of new adaptations, including how they moved, reproduced and responded to their new terrestrial environment. The largest of these ancient amphibian ancestors was the crocodile-like predator *Prionosuchus* from Brazil, which grew to 9 metres long.

During the Mesozoic, reptiles emerged in competition with these giant predatory amphibians, which may have contributed to their decline in abundance, diversity and size. By the middle of the Cretaceous, just one of the giant amphibians survived in the cold river systems of southern Australia, over 100 million years after the last known record of its closest relatives.

Koolasuchus cleelandi may have grown to 5 metres long and weighed as much as half a tonne. Although it may have looked a bit like a giant axolotl, *Koolasuchus* was probably an ambush predator, lurking just beneath the surface of the water waiting to snap up passing fish, turtles and crustaceans, perhaps even a small dinosaur, in its large jaws. Today, the largest living amphibian is the Giant Chinese Salamander, which grows to 1.8 metres in high mountain streams and lakes. Their preference for cold water probably protected these large amphibians from competition with crocodiles. Once the climate began to warm in the later Cretaceous, crocodiles began to spread and diversify, taking over as the dominant large freshwater predator.

Creatures of the air

Fossilised feather embedded in rock from Koonwarra.
Photographer: Frank Coffa, Museums Victoria

Pterosaurs

KINGDOM: Animalia

PHYLUM: Chordata

CLADE: Ornithodira

ORDER: Pterosauria — wing lizard

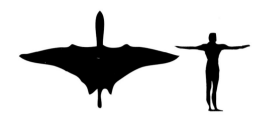

SPECIES	ETYMOLOGY	WINGSPAN
Mythunga camara	Star hunter of the skies	4–5 m
Aussiedraco molnari	Aussie dragon after palaeontologist Ralph Molnar	5 m

The pterosaurs were a distinctive and unusual group of flying reptiles that first appeared in the fossil record 220 million years ago (in the Upper Triassic) and had entirely disappeared by the end of the Cretaceous period, 155 million years later. The earliest pterosaurs were often small animals, about the size of a pigeon, with teeth and long, stiff tails. Later species of the Late Jurassic period often lacked the teeth and tails, and some grew to enormous sizes. They ranged from the tiny sparrow-sized *Nemicolopterus crypticus* from China to the Texan giant *Quetzalcoatlus* which, with a wingspan of 11 metres, was the size of a small aeroplane.

Rather than having the webbed fingers of a bat-wing, pterosaur wings stretched from a single, incredibly elongated finger, allowing them to walk, climb or grasp with the remaining fingers located midway along the wing.

Early reconstructions of pterosaurs assumed that their flight was ungainly and inefficient, but the inherently high-energy demands of flight mean that pterosaurs must have developed an efficient and effective means of flight, not only to survive but also to have radiated into such a spectacular array of forms. The energy demands of flight become even more challenging at both small and large scales, and pterosaurs appear to have flown at sizes as small as, and vastly larger than, modern birds.

Reconstructions of pterosaur flight dynamics suggest that at least the large species were adapted for slow, long-distance flying. Their stiff necks could not be folded (like pelicans or egrets) in flight, but would have been fully extended (like geese or swans). Long, narrow wings are characteristic of birds like albatross, which soar across vast oceanic distances in search of food, with little energy

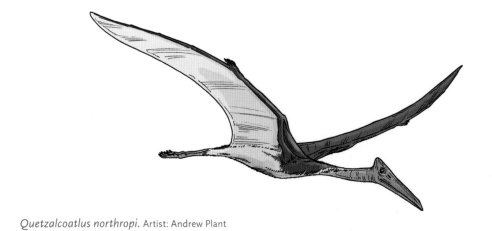

Quetzalcoatlus northropi. Artist: Andrew Plant

expenditure. Pterosaurs may, however, have been highly manoeuvrable, like modern frigate birds, enhancing their ability to catch fish.

The pterosaurs probably walked upright on two legs, although this is a subject of great argument among experts. Some think that the pterosaurs walked on all fours, rather like bats do on the ground. The poor preservation of the fragile pterosaur skeletons makes this debate difficult to resolve, particularly as we cannot be certain if the wings were separated from the pterosaurs' legs (like birds) or attached (like bats). Recent studies of pterosaur footprints clearly show that they could walk on all fours.

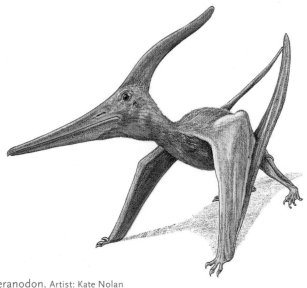

Resting pteranodon. Artist: Kate Nolan

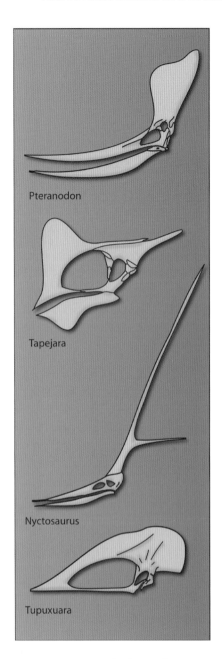

Pteranodon

Tapejara

Nyctosaurus

Tupuxuara

Pterosaurs evolved a startling variety of skull shapes, many of which may have supported ornate decoration.

Only about one hundred species have been discovered around the world. Those that have been described, however, are highly diverse and distinctive. Different species exhibit strikingly different crest and snout shapes. These crests seem to have only developed later in life as a feature of sexual maturity. Many pterosaur species show two distinct size groups of individuals; the larger bodied individuals tend to have larger crests and a narrower pelvis, suggesting that the males may have been larger and more flamboyant than the females. In pteranodontids, the males may have been a third larger than the females, and presumably used their size and large crests to compete with each other to attract females.

Pterosaurs may have aggregated in large numbers at feeding grounds, on roosting cliffs or to compete for mates, but they probably did not have the complex family life often seen in birds. Unlike most birds but like many other reptiles, pterosaurs laid soft, leathery eggs that would have been incubated underground. Young pterosaurs hatched with their wings fully formed and were able to fly almost immediately, suggesting that they were independent from a very early age. While some reptiles guard their nests and care for their young once they hatch, pterosaurs probably did not look after their young for an extended time.

Pterosaur fossils are most often found in coastal deposits, suggesting that they were probably marine specialists much like seabirds today. Some fossils have been found with fish preserved in their stomachs, while one specimen has been found with a collection of fish and crustaceans preserved just below the jaw—suggesting that it may have used a fishing pouch like pelicans do. The teeth of early pterosaurs were ideally suited to catching fish. Many pterosaurs also featured a beak-like

Ornithocheirus sp. with turtle *Notochelone costata* and ammonite. Artist: Frank Knight

hook on their snout, which is also useful for fish-eaters. The later toothless pterosaurs retained many features suggesting a diverse range of specialisations for marine prey: one species had pointed upturned jaws, which might have been used for opening shells. Another species was a specialised wader, sifting the shallow waters for food, while an Argentinean pterosaur used a sieve, like whale baleen, to feed on plankton. Some even moved into the forests where they fed on insects.

Apart from these specialist feeders, pterosaurs are most likely to have foraged for fish from the surface of the water like modern day gulls and skuas. They show none of the specialised skull adaptations for plunge diving found in modern gannets, nor do they have the flexible neck required to catch fish from the wing like modern frigate birds.

Despite their popular representation as gaunt, bald creatures, pterosaurs were just as well muscled as other comparable creatures, despite their light and fragile skeletons. Their extravagant crests were almost certainly brightly coloured and some pterosaur fossils suggest that they were furred. With the high metabolic demands of flight, they were almost certainly warm-blooded, particularly small species or young pterosaurs. Some may even have been nocturnal, as petrels are today. No doubt they made the same noisy, conspicuous contribution to the Eromanga seaside as modern seabirds do today on the Australian coast.

Although pterosaurs were widespread along the coastal and estuarine regions of the inland sea, relatively few of their fossils have been found in Australia. *Mythunga camara*, with a wingspan of over 4 metres, was the first to be formally named. The large, interlocking teeth of this species support the idea that it probably ate relatively large fish. The fragmentary remains of other pterosaurs suggest that several other species may have also occurred in Australia, including *Aussiedraco molnari*, but further species cannot be identified until more complete remains are found.

The Australian pterosaur may well have looked similar to this South American species, *Anhanguera piscator*.
Artist: John Conway

Birds

KINGDOM: Animalia

PHYLUM: Chordata

CLADE: Aves

ORDER: Ornithurae — bird tails

SPECIES	ETYMOLOGY	HEIGHT
Ichthyornis sp.	Fish bird	30 cm
Nanantius eos	'Anti-bird' of dawn	12 cm

Pterosaurs were not the only reptiles to take to the air over the Eromanga Sea. One group of bipedal running reptiles made an advance on the fine fur covering some of their cousins like the pterosaurs. This new coat initially consisted of simple hairlike filaments, but some soon developed a complex vaned structure:

Nanantius eos. Artist: Carol McLean-Carr

feathers. The evolution of feathers provided a whole range of adaptive advantages including excellent insulation and heat regulation, both for themselves and their eggs, as well as adaptations for flight. Before long, rudimentary wings had appeared on the hands of some of these avian theropods and the transition to modern birds was underway.

The avian theropods were already committed to bipedal locomotion, freeing the arms for other uses. They had inherited the potential for lightweight, air-filled bones from their ancestors. They lost their tails, rotated what had once been a dewclaw into an opposed toe for perching and developed beaks alongside teeth. The fine filamentous coating of vaned feathers spread, radiated and specialised to create an unparalleled aeronautical sheath around the reptilian form. Great feathered aerofoils extended from a single finger, while the remaining fingers shrank and eventually disappeared altogether. A massive keeled breastbone anchored the muscles that would power their launch into the air.

By the Aptian–Albian, primitive birds were well developed and had radiated into a wide range of forms. On the shores of inland seas around the world, many varieties filled the niches now occupied by seabirds. Mesozoic bird nests have been found in vast aggregations, probably much like the breeding colonies of today's seabirds. Colonial breeding by seabirds today seems to be a function of the patchiness of shoaling schools of fish in the ocean, making food difficult to find, but almost unlimited once located. Living in groups can help seabirds to find fish at sea, while the density of food available once found is so great that there is very little competition and great aggregations of birds can live together without running out of food.

Two species of seabird have been found from the Eromanga Sea. *Ichthyornis*, from Lightning Ridge, would have resembled a small gull with teeth. Like many modern seabirds, *Ichthyornis* seems to have been widely distributed, being common in Asia and North America. It was similar in form to the large flightless marine divers *Baptornis* and *Hesperornis*, which hunted schooling fish.

By contrast, *Nanantius eos* was much smaller—about the size of a blackbird. This gull-like bird had clawed wings and probably a dinosaur-like head—except for its beak. Found on the coast near Boulia in Queensland, it probably fed on marine invertebrates and fish. Although its nearest relatives were woodland species, a specimen of this species was found in the stomach of an ichthyosaur, demonstrating a marine connection.

Sharks and fish

Ginsu sharks, *Cretoxrhina mantelli*. Artist: Mark Marcuson/University of Nebraska State Museum

Sharks

KINGDOM: Animalia

PHYLUM: Chordata

SUBPHYLUM: Vertebrata

CLASS: Chondrichthyes — cartilage fish

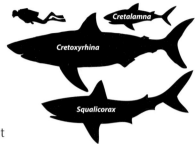

Sharks were an abundant and important part of the marine ecosystem in the Eromanga Sea. Fossilised shark teeth and vertebrae are commonly found throughout the basin. Many of these fossils are of mackerel sharks (Lamniformes), which today include species like makos and tiger sharks.

The inland seas of America were populated with similar species to those of the Australian inland sea, including *Cretoxyrhina* and *Squalicorax* sharks. Artist: Robert Nicholls

Mackeral sharks

ORDER: Lamniformes — thin armour

SPECIES	ETYMOLOGY	TOOTH SIZE
Cretoxyrhina mantelli ('Ginsu' shark)	Cretaceous 'jaws' after paleontologist Gideon Mantell	15.5 mm
Cretalamna sp.	Cretaceous mackerel shark	25 mm
Paraisurus macrorhiza	Related to the mako sharks	18 mm
Carcharias striatula (Sand tigershark)	Banded saw	10.2 mm
Mitsukurinidae sp. (Goblin shark)	After Keigo Mitsukuri	10–20 mm
Archaeolamna spp.	Ancient mackerel shark	20 mm
Echinorhinus australis (Bramble shark)	Southern spiny nose	6 mm
Microcorax sp.	Small crow shark	4 mm
Squalicorax primaevus	Primitive rough crow shark	7.5 mm
Leptostyrax	Delicate lance	18 mm

One of the most impressive of the mackerel sharks was *Cretoxyrhina*, the giant Ginsu shark, nicknamed after a knife for its technique of slicing and dicing its prey. The Ginsu shark is also known from the Western Interior Seaway of North America. It was the largest shark of the time, growing to over 6 metres and regularly feeding on elasmosaurs, fish and giant turtles.

The most common sharks of the Late Albian in Australia were the 'crow sharks', including *Pseudocorax australis*, the much rarer *Microcorax* (known only from a single tooth) and *Squalicorax primaevus*, whose teeth are similar to those of a modern tiger shark. *Squalicorax*, which grew to 5 metres, may well have been a scavenger or opportunistic hunter, as the decayed remains of both marine and land animals have been found associated with its teeth.

Another common shark, a *Cretalamna* species, is typically found in marine deposits, but has also been found in freshwater locations, suggesting that it was a highly adaptable generalist capable of surviving in a wide range of aquatic environments. This large shark grew to 3 metres and seemed to engage in similar feeding frenzies to some modern sharks—one plesiosaur fossil was discovered with eighty *Cretalamna* teeth, from at least seven different sharks.

Sand tiger sharks are the survivors of a genus that once contained numerous species in the Cretaceous and Palaeogene oceans. *Carcharias striatula*, which lived in the Eromanga Sea, was probably quite similar to the modern grey nurse shark, with its elongated tail and smooth pointed teeth for hunting a wide range of small aquatic species. It is likely that this extinct species also reproduced slowly, like its modern counterparts, birthing one or two large live young every couple of years. One of the most unusual sharks was the goblin shark *Mitsukurinidae*. Modern goblin sharks are remarkable for being able to shoot their jaws forward suddenly to strike their prey.

Carpet sharks

ORDER: Orectolobiformes — stretched lobe

Orectolobus maculates, Spotted Wobbegong. Illustration © R. Swainston/anima.net

The carpet sharks (Orectolobiformes) are also unusual-looking sharks—the best known of which today is the wobbegong. These flattened sharks spend their days resting, often in disguise, on the sea floor, but at night become active hunters. Remains of this diverse and distinctive group of sharks have been found from the Eromanga Sea but have not been identified to individual species. Their presence in the Late Albian suggests that at least some areas of the sea floor were well oxygenated.

Squaliform sharks

ORDER: Squaliformes — Primitive type

SPECIES	ETYMOLOGY	TOOTH SIZE
Pristiophorus tumidens (saw shark)	Swollen toothed saw bearer	9–16 mm
Notorynchus aptiensis (cow shark)	Back snout from the Aptian	3–6 mm
Paraorthacodus sp.	Beside the Orthacods	9 mm

 The Eromanga Sea was also home to a broad range of squaliform sharks, which include dogfish sharks, cow sharks, saw sharks and angel sharks. The saw shark, *Pristiophorus tumidens*, used its long flattened snout edged with alternating large and small teeth to slash and disable prey. This species, like modern sawfish, may

Cretaceous sharks' teeth: *Cretoxyrhinidae* (top), *Paraorthocodus* (middle) and *Notorynchus aptiensis* (bottom). Courtesy of Mikael Siversson

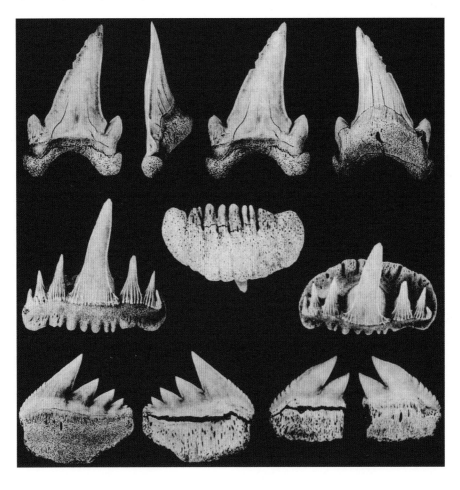

have fed on fish, squid and crustaceans on the sea floor, in waters deeper than 40 metres. These Australian specimens are the earliest known occurrence of this species and the only known occurrence in the southern hemisphere from the time.

The cow sharks have an additional pair of gill slits and are regarded as one of the more primitive forms of living sharks. Today, they are represented in Australia by the broad-nose seven-gill shark (*Notorynchus cepedianus*), a long-lived, large species that cruises the sea floor in deeper coastal waters of the south-east. Like all cow sharks, the seven-gill gives birth to as many as eighty live young, in a shallow estuarine nursery where they stay for a couple of years before moving out into the oceans. In the Eromanga Sea, *Notorynchus aptiensis* was commonly found in sediments from the ancient continental margin.

Ghost sharks

ORDER: Chimaeriformes — Chimera (hybrid creature from Greek mythology)

SPECIES	ETYMOLOGY	LENGTH
Ptykoptychion tayyo	Folded plate, made of stone	1 m
Ptykoptychion wadeae	Folded plate, for Dr Mary Wade	Over 1 m
Edaphodon eyrensis	'Ratfish' from Lake Eyre	1.5 m

The chimaerids are much more distant cousins in the shark family. But, like many sharks, the modern chimaerid sharks appear to have changed relatively little from their ancient relatives. A fossil tooth plate found in Western Australia looks very similar to that of the living Australian ghost shark (*Callorhinchus milii*). These may be the same as tooth plates and fin spines found in Queensland of *Ptykoptychion tayyo*. The ghost sharks were thought to have predominantly eaten shellfish, although larger individuals might have eaten fish as well. Another species, known as a ratfish, *Edaphodon eyrensis*, has been found near Lake Eyre in South Australia and is the earliest known representative of this genus. This long-tailed fish was, like most chimaerids, almost certainly a bottom-feeder on hard-shelled prey.

Opposite: Cladocyclus pankowskii hunting *Diplomystus* sp. and undescribed Lepidotes.
Artist: Brian Engh

Fish

KINGDOM: Animalia

PHYLUM: Chordata

SUBPHYLUM: Vertebrata

CLASS: Osteichthyes — bony fish

The bony fish gave rise to two major groups—the ray-finned fish and the lobe-finned fish. The ray-finned fish include the teleosts, which comprise nearly all of the fish that inhabit today's oceans. Lobe-finned fish include the lungfish, which is capable of breathing air, but more remarkably, lobe-finned fish gave rise to the ancestors of land vertebrates. Together, the descendants of the bony fish dominate both land and water today.

Marine fish

CLASS: Actinopterygii — ray finned fish

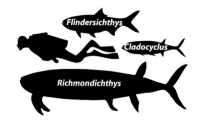

SPECIES	ETYMOLOGY	LENGTH
Richmondichthys sweeti	Fish from Richmond QLD, after Walter Sweet	More than 1.6 m
Australopachycormus hurleyi	Southern pachycormid, after Tom Hurley	More than 1 m
Cooyoo australis	Southern fish (Pooroga language)	3 m
Cladocyclus geddesi	After discoverer and preparator Kerry Geddes	80 cm
Pachyrhizodus marathonensis	Thick-rooted tooth, from Marathon, QLD	Less than 80 cm
Pachyrhizodus grawi	Thick-rooted tooth, named for Booree Park station manager 'Beno' Graw	45 cm
Dugaldia emmilta	From the red-tinged limestone of Dugald River, QLD	40 cm
Flindersichthys denmeadi	Flinders fish, after geologist Alan Denmead	1.25 m
Euroka dunravensensis	For the Euroka Arch formation and Dunraven station, Hughenden QLD	More than 1 m

Cooyoo australis
Artist: Paul Stumkat

Most fish fossils are those of fast-swimming predators. The largest of these was *Cooyoo australis*, which grew to 3 metres long. *Flindersichthys denmeadi* was slightly smaller and also carnivorous, as evidenced by a fossil with smaller fish preserved in its mouth. This abundant species is commonly fossilised in the northern regions of the inland sea, close to the area where water moved in and out from the northern Carpentaria Basin.

Pachyrhizodus marathonensis, another large, fast-swimming, predatory fish, is also common in the inflow regions for the inland sea, as was its smaller relative, *Pachyrhizodus grawi*. Such inflow regions may well have provided a rich supply of food. Other carnivores included the metre-long, swordfish-like *Australopachycormus hurleyi*, described as a hyper carnivore with enlarged fangs.

By contrast, *Richmondichthys sweeti* was a gentle giant, exceeding 1.6 metres long. *Richmodichthys* had no teeth and fed by gulping large mouthfuls of water and filtering out plankton. This common, slow-moving creature might have been at the mercy of the predatory reptiles and sharks in the Eromanga Sea, but for its thick armour of bony scales. Also slow-moving was the eel-like *Euroka dunravenensis* which may have concealed itself in crevices to ambush prey.

Many of these large fish are also found in small, presumably juvenile forms, perhaps co-existing with other smaller species like *Dugaldia emmilta*, which grew to little more than 40 centimetres in length.

Other fish have also adapted to freshwater environments. While *Cladocyclus*, a genus of fast-swimming, open-water predators, are

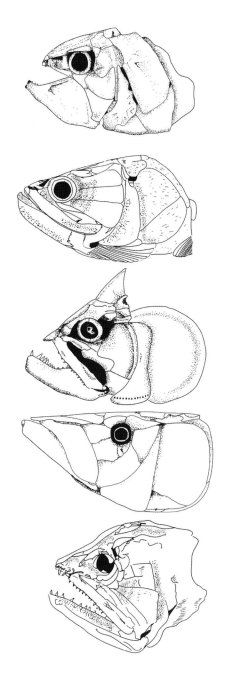

Fish skulls (from top):
Dugalida emmilta,
Flindersichthys denmeadi,
Coyoo australis,
Richmondichthys sweeti and
Pachyrhizodus marathonensis
(not to scale)

Model of *Pachyrhizodus caninus*. Photo courtesy Triebold Paleontology Inc., Woodland Park, Colorado

most common in shallow marine or estuarine deposits, the Australian species *Cladocyclus geddesi* was unusually found in freshwater deposits, suggesting that this species was capable of moving upstream in search of prey.

Freshwater fish

CLASS: Actinopterygii — ray finned fish

SPECIES	ETYMOLOGY	LENGTH
Leptolepis koonwarri	With delicate scales, from Koonwarra	8 cm
Leptolepis crassicaudata	With delicate scales and a thick tail	7 cm
Coccolepis woodwardi	With rough scales, after palaeontologist Arthur Smith Woodward	11 cm
Wadeichthys oxyops	Fish with a 'spike cheek', named for Rev. RT Wade	9.5 cm
Koonwarria manifrons	From Koonwarra with hand-like bone structure	18 cm
Psilichthys selwyni	Without scales, after geologist Alfred Selwyn	50 cm

The Koonwarra fish beds were discovered in 1961 when road workers rebuilt a section of the South Gippsland Highway. This rich collection of small fish and invertebrates provides a rare insight into the ecology of the freshwater streams that may have run into the Eromanga and surrounding seas. The large number of small fish preserved at Koonwarra suggest a mass fish kill—perhaps due to a sudden lack of oxygen as might be caused by ice, or a layer of volcanic ash.

Some fish from the Koonwarra Formation had a distinctly primitive appearance. *Coccolepis woodwardi* was covered in diamond-shaped scales, rather than the rounded scales of many modern fish, and had a long robust jaw, similar to those seen in earlier Devonian fish. Other fish had a more modern appearance. *Leptolepis* was one of the very first modern teleost fish, characterised by their highly modified tail, which provided a powerful boost to fast swimming. *Leptolepis* looked a bit like modern herring, and lived in large schools, feeding on surface plankton. Although *Leptolepis* were not actually close relatives of herrings, *Koonwarria manifrons* does belong to the same important modern family of small fish that includes anchovies, pilchards, sardines and other fish, which comprise a large part of commercial fisheries.

Leptolepidae fossil found near Koonwarra. Photographer: Benjamin Healley. Museums Victoria

Fossil of *Wadeichthys oxyop*, found at Koonwarra. Photographer: Benjamin Healley. Museums Victoria

While many of these fish families are found worldwide, *Wadeichthys oxyop* was found only in Australia. Other freshwater fish of the Early Cretaceous included the catfish, *Psilichthys selwyni*, and fossils that appear to resemble an early eel.

Lungfish

CLASS: Sarcopterygii — lobe-finned fish

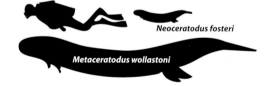

Neoceratodus fosteri

Metaceratodus wollastoni

SPECIES	ETYMOLOGY	LENGTH
Ceratodus nargun	Horny tooth, rock monster (Indigenous)	1 m
Neoceratodus forsteri	New Ceratodus after William Forster	1 m
Metaceratodus palmeri	After Ceratodus, after Colonial Secretary Arthur Hunter Palmer, QLD	2 m
Metaceratodus wollastoni	After Ceratodus, after collector and opal dealer Tullie Wollaston, SA	4 m
Archaeceratodus avus	Original Ceratodus, grandfather	1 m

Lungfish, *Neoceratodus gregoryi*. Artist: Frank Knight

Lungfish belong to an ancient class of freshwater fish, the Sarcopterygii—the lobe-finned fish. These fish were common in the Devonian Seas of around 400 million years ago. Today only six species of lungfish survive. Lungfish can breathe air through simple lungs, which allows them to survive when water dries out during droughts. They are also able to burrow into the mud and aestivate, or become dormant, during droughts. They are long-lived survivors, both as a species, and as individuals, living for over eighty years.

The only living Australian species, the Queensland lungfish *Neoceratodus forsteri*, seems to have changed little in over 100 million years. Fossils identical to this living form have been found in the Griman Creek Formation, near Lightning Ridge, on the edge of the Eromanga Sea. Another lungfish, *Archaeceratodus avus*, was similarly long-lived. Its fossils have been found in Triassic Hawkesbury Sandstone as well as the Albian Eumeralla Formation, separated by 120 million years. There are several other fossil species of lungfish known from the lakes and rivers that might once have fed into the Eromanga Sea, including one similar in size to modern lungfish, *Ceratodus nargun*. Other species were much larger than modern lungfish, including *Metaceratodus palmeri*, and a giant lungfish, *Metaceratodus wollastoni*, which may have grown up to 4 metres long. Its fossils are often opalised and have even been found in marine deposits, suggesting this freshwater species was sometimes washed into the salty waters of the inland sea.

Lungfish, *Neoceratodus forsteri*, at Melbourne Museum.
Photographer: Benjamin Healley. Museums Victoria

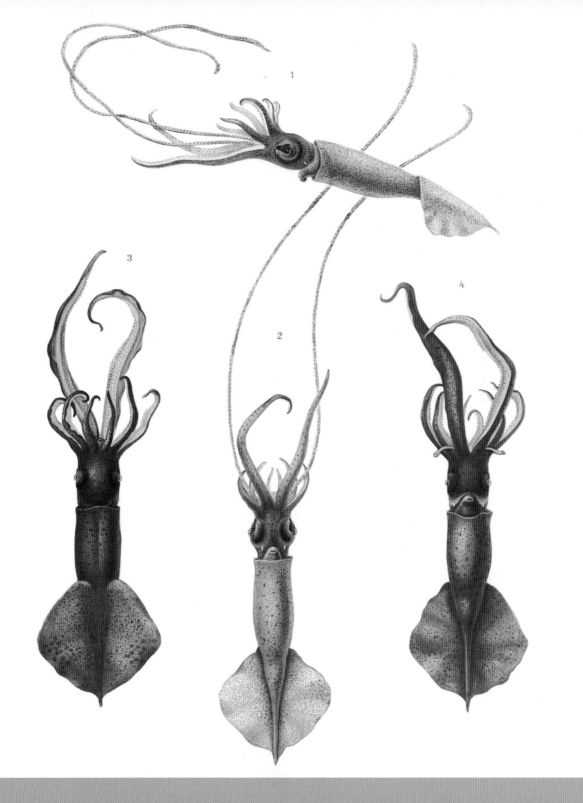

Molluscs

Images of squid from *Die Cephalopoden* by Carl von Chun. Published by G.Fischer in 1910.

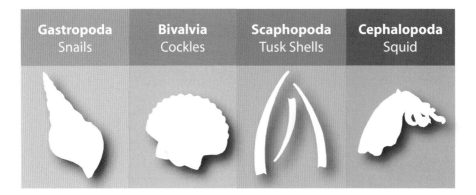

Gastropoda Snails	Bivalvia Cockles	Scaphopoda Tusk Shells	Cephalopoda Squid

KINGDOM: Animalia

SUPERPHYLUM: Lophotrochozoa

PHYLUM: *Mollusca* — thin-shelled

Molluscs are one of the most ancient groups of animals on Earth. Mollusc-like creatures first appear in the fossil record more than 540 million years ago, in the very earliest records of complex multicellular life. This ancient and diverse phylum of invertebrates ranges from microscopic shells too small to see with the naked eye to giant squid over 10 metres long.

Shellfish

Gastropods

CLASS: *Gastropoda* — mouth foot

GENERA

Anchura	*Cellana*	*Pleurotomaria*
Ampullina	*Diodora*	*Vanikoropsis*
Avellana	*Euspira*	

The best-known gastropod is the garden snail, but they are more commonly marine species characterised by spiral shells (like snails) or cap-shaped shells (like limpets). In the Aptian, the inland sea floor teemed with a great variety of gastropods. Sixty-three different species have been recorded so far.

Some of them were similar to modern species. The moon shell *Euspira reflecta* probably hunted by engulfing other shells with their large foot and drilling neat

The Cretaceous moon shell *Euspira reflecta* was a common predator, much like modern members of the family such as this nocturnal East Timorese species *Naticarius orientalis*
Photographer: Nick Hobgood

holes into them, just as they do today. Moon snails (also known as necklace shells) are also responsible for the delicate sand collars in which they lay their eggs. By contrast, the similar *Ampullina* have no living relatives today.

Other gastropods were grazing herbivores, such as *Vanikoropsis* and *Pleurotomaria*, which fed on sponges. *Pleurotomaria* is one of the oldest groups of gastropods but was all but wiped out in the Cretaceous mass extinction. Modern varieties of these shells are now restricted to deep waters but were once widespread in the shallow inland seas. Other grazers included the limpets, like *Diodora* and *Cellana*, which used their powerful foot to suck themselves onto the rocks beneath their shells like conical hats, protecting themselves from wave action and predators on the seashore.

Tusk shells

CLASS: Scaphopoda

GENUS: *Dentalium* — tooth-like

A few species of *Dentalium* tusk shells inhabited the Aptian sea floor. These gently curved tubular molluscs dig into the sea floor, leaving one end sticking out at an angle to filter feed. Although they are generally found today in deeper water, the shells of these molluscs are popular for decoration and exchange among the Indigenous people who harvest them from the sand of Kyuquot Sound near Vancouver Island in North America.

Tusk shell, *Dentalium intercalatum*. Photographer: Karen Gowlett-Holmes

Bivalves

CLASS: Bivalvia — two valves

GENERA

Apiotrigonia	Indogrammatodon	Opisthotrigonia
Astarte	Inoceramus	Panopea
Aucellina	Iotrigonia	Phaenodesmia
Campionectes	Laevitrigonia	Pseudavicula
Camptochlamys	Limea	Pterotrigonia
Camptonectes	Maccoyella	Syncyclonema
Cyrenopsis	Maranoana	Tancretella
Ennucula	Meleagrinella	Tatella
Eyrena	Myophorella	Tellina
Fissilunula	Nanonavis	Trigonia
Glycimeris	Nototrigonia	Venericardia
Grammatodon	Onestia	Yoldia

Bivalves, commonly represented as cockles or clams, were by far the most common mollusc of the Eromanga sea floor. In the Aptian, isolation and a rapidly cooled climate seem to have driven rapid change in the molluscs and many species in the Eromanga Sea were unique to the Australian region, including the mussel, *Eyrena linguloides*, and the fan shells *Maccoyella reflecta*.

As the Aptian gave way to the Albian the Eromanga sea floor seems to have changed, from a structurally diverse habitat capable of supporting a wide array of species, to a sediment-laden basin, dominated by just a few species. The great rivers washing into the Eromanga slowly covered the sea floor with sediment. Geological activity seems to have made the inland seabed drop at this time. This reduced the impact of waves and cold ocean currents on the sea floor, making the waters both warmer and less oxygenated.

The Albian seafloor was crowded with filter-feeding bivalves. Photographer: Danielle Clode. Source: South Australian Museum

Only a few species of the strikingly sculptural trigonia family, common and widespread in the Aptian, survive today in deep water off Australia. Photographer: Benjamin Healley, Museums Victoria

By the Albian era, there were fewer species of shells (only forty-three species are known), but some of those occurred in great numbers. Buried just below the surface lay an abundance of life—thousands upon thousands of filter-feeding bivalves, thrusting their long siphons up out of the sand to sift microorganisms from the water.

The giant mussel *Inoceramus sutherlandi* (growing up to 50 centimetres long) was so common that today it dominates the shelly layer of the striped Toolebuc Formation—the fossilised sea floor that stretches across inland Queensland, South Australia and the Northern Territory. *Inoceramus* seems to have had a slightly flexible shell, rather than the hard brittle shells of modern bivalves: an adaptation to resist breakage by predators. The clam-like shells of *Aucellina hughendensis* were also common.

This poor environment with less oxygen and deeper warmer water may not have suited many species, allowing a few resilient filter-feeders like *Inoceramus* to dominate the sea floor. Higher up in the water column, however, the seas remained rich and healthy. And a diversity of remarkable molluscs flourished here—the free-swimming cephalopods.

Cephalopods

CLASS: Cephalopoda — head foot

The word cephalopod comes from the ancient Greek words for head and foot, an apt name for a creature whose limbs or tentacles surround its mouth. Today we know the cephalopods primarily from squid and octopus species, but the prehistoric seas teemed with a great diversity of ancient cephalopods very different from those we know today.

All cephalopods are exclusively marine creatures—they are never found in freshwater. And while they are more common in tropical seas than in polar regions, cephalopods can be found in every layer of the ocean, from the surface to the deepest abyssal plain.

The cephalopods have diversified greatly from their molluscan ancestors. While some groups (like the ammonites and nautiloids) retained their external shells, others (like the coleoid squids and octopus) have internalised their shells, or lost them entirely. Like other molluscs, cephalopods are blue-blooded, using a copper-containing protein to transport oxygen rather than iron-red haemoglobin. But their nervous systems are much more complex than those of other molluscs. With their large nerve cells, advanced eyesight and big brains, cephalopods display a remarkably high level of intelligence and responsiveness to their environment.

Most of the cephalopods move by jet propulsion—sucking water into an internal muscular sac that contains their gills and expelling it through their constricted siphon. They are commonly fast-moving hunters, capable of concealing themselves in a cloud of ink or camouflaging themselves with changing skin colours. These skin colours are also used as a means of communication between individuals.

Nautilus

SUBCLASS: Nautiloidea — sailor

SPECIES

Cimomia sp.

Kummelonautilus hendersoni

Of all the cephalopods, the living species of nautilus look the most like typical molluscan shellfish, being able to fully retreat into their large external shells.

These animals alter the gas and liquid in the chambered sections of their shells to regulate their buoyancy. These chambered shells are often cut in half to display their beautiful mother-of-pearl interiors.

A *Cimomia* nautilus
Photographer: Rodney Start, Museums Victoria

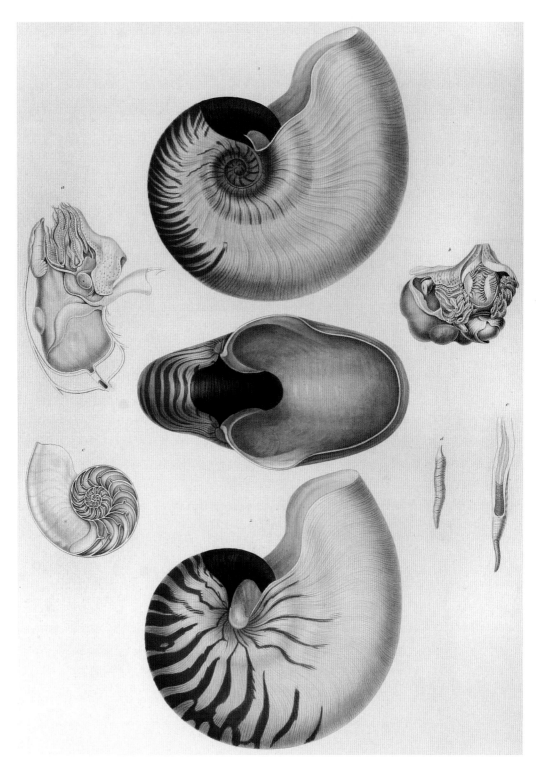

The living species of nautilus suggest that their Cretaceous relatives may also have been beautifully coloured and patterned. Source: From *Illustrations of nautiluses* by Jean-Charles Chen, 1842.

There are only six species of nautilus living today, most of which are found in waters between 150 and 300 metres in depth. They cannot survive long in waters warmer than 25° Celsius nor at pressures greater than those at around 800 metres.

Nautilus differ from other cephalopods in that they lack both an ink sac and advanced vision. These animals are still relatively intelligent for molluscs, but lack the advanced problem-solving and memory capacity of modern octopuses. Nor are they such speedy or manoeuvrable swimmers. Unlike other modern cephalopods, nautilus species are relatively long lived and slow breeding, producing just a few large eggs at a time, with their young hatching at a relatively large size. These slow-breeding characteristics are common to species living in less productive deep-sea environments.

In the past there were many more nautilus species than today, exhibiting a much greater variety of shapes. In the Palaeozoic record some had elongated straight shells (orthocones), others were gently curved (cyrtocones), while some had loose open coils (gyrocones) or even bulbous spindle shapes (brevicones). Some species of orthocones were truly giant predators, growing up to 5 metres long.

The nautiloid species from the Eromanga Sea, however, were similar in form and size to the modern nautiloids and usually less than 20 centimetres in diameter. *Cimomia* was a smaller species found in Western Australia, whereas *Kummelonautilus hendersoni* was a larger, more common species found in Queensland.

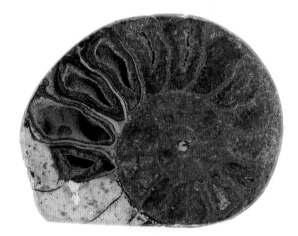

Cretaceous ammonite fossil from the Gulf of Carpentaria. Photographer: Mary-Anne Binnie, South Australian Museum

Ammonites

SUBCLASS: Ammonoidea —

like the ram's horn of the Egyptian god Ammon

Tropaeum imperator

GENERA

Acanthoceras	*Hamites*	*Myloceras*
Anisoceras	*Hypoturrilites*	*Naramoceras*
Australiceras	*Hysteroceras*	*Pseudoheliceras*
Beudanticeras	*Idanoceras*	*Ptychoceras*
Desmoceras	*Idiohamites*	*Worthoceras*
Goodhallites		

Ammonites are one of the most widespread, common and diverse groups present in Cretaceous seas. They rapidly evolved into so many numerous, distinctive and abundant species, that their presence is used to identify the different geological ages of the Cretaceous.

Ammonites probably produced large numbers of very small eggs, suggesting that they may have been fast-growing and short-lived like most modern squid and octopus. Some species were sexually dimorphic with males and females of different sizes, allowing larger females to produce larger egg clutches.

Ammonites varied in size from the tiny *Faciferella*, just 2.5 centimetres across, to the massive *Parapuzosia seppenradensis*, an uncoiled ammonite, which grew to over 3 metres long. In Australia, the largest ammonite was *Tropaeum imperator*, reaching 75 centimetres across, which was mistaken for a tractor tyre when first discovered.

Early ammonites followed the typical tightly coiled symmetrical body plan characteristic of modern nautilus. During the mid Cretaceous, however, the ammonites diversified into a wide

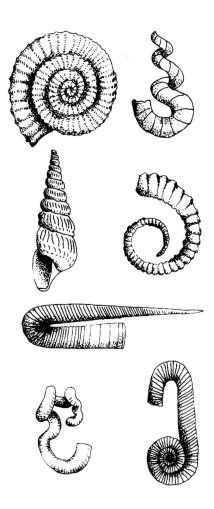

Ammonite shells evolved a great diversity of different forms.

Tropaeum imperator, Giant ammonite.
Photographer: Danielle Clode. Source: South Australian Museum

range of highly divergent body plans, including straight, tubular and hooked forms, to those with corkscrew spirals.

Many Aptian ammonites retained the traditional tightly spiralled body plan, such as members of the Opellidae family, which were small ammonites typically around 5 centimetres across. The Ancyloceratidae family also retained their spiral form, but the spirals are often detached and heavily ribbed.

In the Albian the ammonites diversified greatly. Some ammonites retained the conservative body plan (such as the Desmoceratidae, Opalinidae and Brancoceratidae) perhaps because they were fast-swimming plankton feeders. One of the most common of these ammonites was *Goodhallites goodhalli*, a large, heavily ribbed ammonite with delicate fractal patterns on the complex surfaces between the internal shell chambers.

The Anisoceratidae (including *Anisoceras*, *Laberceras* and *Myloceras*) lived within a hook-shaped space in their shells, which had a loose spiral or helical shape, often ornamented with ribs and spines on the outside, and growing

Desmoceras cardensis, ammonites from the Northern Territory.
Photographer: Alexis Tindall, South Australian Museum

to around 20 centimetres. The diversity of these uniquely Australian species may have prevented colonisation of the inland sea by widespread species from elsewhere.

The Baculitidae (such as *Lechites gaudini*) had a long, straight, cone-shaped shell, similar to the orthocone nautiloids. These shells lacked a counterweight in their tip, suggesting that the animals may have floated head down in the water column, perhaps only shifting to a horizontal plane to dart forwards at speed. Their shells are often found in large numbers, suggesting that they may have lived in shoals, possibly as plankton feeders. Hamitidae of this age had a similar loose spiral or helical shape to Anisoceras, but without the spines and ridges; they commonly featured simple ribbing on their shells. Characteristic Australian species in this group include *Hamites venetzianus* and *Hamites virgulatus*.

The ptychoceratid ammonites look a bit like a paperclip, with a single straight shaft folded back upon itself. Like the Hamitidae, these ammonites were also sometimes ribbed, although only weakly, whereas others were smooth. Australian species include *Ptychoceras adpressum* and *Ptychoceras closteroides*.

Scaphitid ammonites retained a tight coil for much of the shell but the last part forms a distinctive hook with characteristic suture lines and ribbing. Scaphitidae started their growth in a tight coil, the final hook marking sexual maturity. This final part of the shell might have served as a brood chamber for eggs, although it is present in both male and female shells. Australian Scaphitidae include *Worthoceras worthense* and *Scaphites eruciformis*.

Artist's impression of a living turrilitid ammonite.

The Turrilitidae were also coiled, although in a helical corkscrew pattern like a snail. This shape would not allow them to swim, so they are assumed to have lived on the sea floor, like modern gastropods, using their tentacles to search for food. Other reconstructions, however, have the tentacles pointing upwards, and suggest that these ammonites may have gently 'corkscrewed' through the water column, sifting plankton with their widespread tentacles. Australian Turrilitidae include *Mariella bergeri*, *Notostrepites exilis*, *Pseudohelicoceras gracile* and *Pseudohelicoceras catenatus*.

Belemnites

SUBCLASS: Belemnoidea — dart-like

Peratobelus oxys

SPECIES

Peratobelus oxys

Peratobelus australis bauhinianus

Dimitobelus diptychus

Dimitobelus stimulus

Dimitobelus dayi

Belemnites looked similar to modern squid; they were elongated animals with ten arms around a central beak, large eyes and an ink sac. Unlike squid, however, their conical bodies were supported internally by a calcified guard, fossilised as a long bullet-shaped shell. Large deposits of these guards have been dubbed 'belemnite battlefields' and perhaps formed when the belemnites died *en masse* after breeding. The hard guard formed only a small part of the animal. The guard of one European belemnoid, *Megatheuthis gigantean*, measures 45 centimetres long, which gives the animal an estimated total length of 3 metres.

Like modern squid, belemnites were active hunters, grabbing their prey with hooked arms. These hooks are often found in the fossilised stomachs of ichthyosaurs, suggesting belemnites were the primary diet of these much larger predators.

During the Aptian period, the Eromanga Sea was home to just two species of belemnite, both of which were widespread. *Peratobelus oxys* was a large belemnite, estimated to have grown to 72 centimetres long, with guards up to 12 centimetres, and more than twenty annual growth rings. *Peratobelus australis bauhinianus* was slightly smaller and more slender.

Belemnites were more diverse in the Albian and represented by species of *Dimitobelus*, which had a

Opalised belemnite cephalopod, *Neohibolites eromos*.
Photographer: Rodney Start, Museums Victoria

characteristic southern distribution ranging across Australia, West Antarctica, South America, New Zealand and New Guinea. They seem to have been relatively long lived as their guards reveal up to 10–12 growth rings. *Dimitobelus* includes some large and widespread species, *Dimitobelus diptychus* and *Dimitobelus stimulus*, both with guards up to 10 centimetres in length, suggesting the living animals may have been 50 centimetres long. The medium-sized *Dimitobelus dayi* was more common in the northern waters of the Eromanga Sea in the Early Albian.

All that remains of the soft-bodied belemnites is their hard internal 'guard', rather like the modern cuttlefish 'bone' commonly found on beaches. Photographer: Rodney Start, Museums Victoria

Coleoidea

SUBCLASS: Coleoidea — sheath-like

Boreopeltis soniae

SPECIES

Boreopeltis soniae   *Trachyteuthis willis*

Muensterella tonii

The Coleoidea include over eight hundred living species of octopus, squid and cuttlefish today. They have either eight or ten tentacles and are typically short-lived, fast-growing species. Most lay large clutches of eggs and die shortly afterwards. Their most famous modern members are the deep-sea giant squid, which grow up to 14 metres long.

The giant vampiropod squid *Boreopeltis soniae* may have been a significant predator in the inland sea. Artist: Janet Fickling, Kronosaurus Korner.

Large squid were also a feature of the Eromanga Sea. A giant species, *Boreopeltis soniae*, was recently found near Richmond in Queensland by fourteen-year-old Sonia Levers. The internal gladius or squid pen of this specimen was 1.3 metres long. This would suggest the complete animal was just over 1.5 metres in length. Another smaller species was discovered at the same location as well as *Meunsterella tonii*, which belongs to the same family as the modern giant squid.

The soft-bodied coleoids leave few fossils so it is difficult to know what these three ancient species looked like. They would not have been adapted for the same deep-water environment as modern giant squid. Similar giant species in the American Western Interior Seaway were vulnerable to predation by mosasaurs and large fish.

But the greatest risk to the cephalopods, like all other residents of Australia's inland sea, was not predation. The climate was changing; sea levels were falling. Soon the great lineages of cephalopods that dominated the Eromanga Sea, in abundance and diversity, would all but disappear, along with the inland sea itself.

Modern squid range in size from this tiny Southern Bobtail Squid, *Euprymna tasmanica*, to the legendary deep-sea giant squid. Photographer: Julian Finn, Museums Victoria

Guard of *Boreopeltis soniae*, vampiropod squid, measuring 106 cm. Photographer, Timothy Holland, Kronosaurus Korner

The end of the world

The disappearing ocean

The crater at Wolfe Creek, Western Australia, is an example of the impact of a relatively small meteor strike. Photographer: Dermot Henry, Museums Victoria

Stretching out from the edge of the granite escarpment lies an ocean that has receded from all time and memory. The waters have departed; the great beasts vanished. The fish, the shells, the reptiles—large and small—all have disappeared. Where once there were shallow fertile seas, a vast empty plain stretches as far as the eye can see, glittering in the heat, and the darkened lines of creeks and rivers meander their way across the wide flat interior of north-western Cape York Peninsula.

As the seawaters warmed, the cold-water specialists declined. Globally, the Late Cretaceous was the warmest period known in Earth's living history. By the end of the Albian, the ichthyosaurs disappeared, making way for mosasaurs. These marine reptiles not only dominated the warmer seas surrounding Australia, but also infiltrated freshwater habitats, with their relatives ultimately giving rise to the lizards and snakes of modern Australia. In the warmer climate, the giant amphibious predators finally gave way to the crocodiles. Generations of ammonites and belemnites continued to speciate, diversify, radiate and disappear.

With the passing of the Albian, the great inland seaway retreated to the north, to the Gulf of Carpentaria, and to the south a growing rift emerged between the Antarctic and Australian continents. Soon, only Tasmania provided a bridge between them, but in the last age of the Cretaceous, even that link was periodically severed. Australia began its long, lonely drift north, while the cold Southern Ocean encircled Antarctica, eventually over time beginning a long slow freeze that ultimately smothered its ancient forests under a thick blanket of ice and snow.

The impact of meteors

But it was not landmasses drifting in oceans, but landmasses drifting in space that ultimately ended the Cretaceous. Meteors had struck Earth before. On the edge of the inland sea, near Oodnadatta, a meteor created a crater 3–4 kilometres across, now known as Mount Toondina. Buried deep within the rocks of the Eromanga Basin in south-west Queensland lie the twin craters of Tookoonooka and Talundilly, created by meteors that struck in the Mid to Late Aptian. Later still, in the Late Cretaceous, a meteor strike created the Yallalie impact structure in the Perth Basin.

These massive impacts, devastating as they must have been, were nothing compared to the asteroid that struck on the other side of the globe, marking the end of the Cretaceous. When the Chicxulub asteroid, 10 kilometres across, struck off the coast of Mexico, it created a crater more than 180 kilometres across. The impact released energy the equivalent of 100 000 000 megatons of TNT. By comparison, the Hiroshima nuclear bomb released just 16 megatons while the volcano of Krakatoa in 1883 released 200 megatons. The eruption of Krakatoa in Indonesia could be heard 3000 kilometres away in Perth and Mauritius. So much debris was thrown into the atmosphere that the global climate was cooled and disrupted for several years, creating a summerless Europe.

Chicxulub created a summerless world.

So much debris was thrown into the air that the world turned dark. Sunlight could not penetrate. Plants and plankton could not photosynthesise. In the depths of the oceans, where sunlight never reached, in the streams and the forests, the detrivores, scavengers and opportunists scraped by on the remains of a once rich world, waiting for the sun to return. Around the globe a layer of clay containing high levels of iridium, an element rare on Earth but common on asteroids, left a line in the sand between the age of dinosaurs and the age of mammals. Mega tsunamis swept the world's oceans and increased volcanic activity added to the Earth's woes.

A mass extinction

Over the course of many millions of years, three-quarters of the species known on Earth became extinct. The giant dinosaurs disappeared, their

Changing sea levels over time

Future...

Neogene

Palaeogene

ALBIAN

APTIAN

Cretaceous

Jurassic

Triassic

Permian

Carboniferous

Devonian

Silurian

Ordovician

Cambrian

surviving ancestors spreading their feathered wings and taking to the air as birds. The oceans were particularly devastated. The primary producers, who turned sunlight into food, died rapidly. The upper level filter feeders, dependent on plankton, starved. The great reef-building corals disappeared. The ammonites and belemnites ceased their proliferation, the last of their kind layering the sea floor with their elegant remains. Sharks, fish and other molluscs suffered great losses. Many of the large marine reptiles that fed on the filter feeders ran out of food and disappeared.

Over time, a new cycle of evolution began. A new world, with new oceans— without plesiosaurs and ichthyosaurs, without pterosaurs and ammonites. A new world, in which mammals would join the survivors, the turtles, sharks and fish, to repopulate the Earth. A new world, in which one fledgling species, in the course of just 100 000 years, would spread across the globe, changing the face of the planet, restructuring entire ecosystems, driving thousands of species to extinction and expelling their own greenhouse gases into the air, driving a pattern of climate change and sea-level rises the consequences of which are as yet unknown.

As we sit on the edge of the escarpment, looking out over a timeless, yet ever-changing landscape, we can only wonder what kind of world awaits beyond the age of humanity. Will the inland sea emerge once more in the dry heart of Australia? And will there be anyone here to see it?

GLOSSARY

aeronautical sheath / aerofoils structures that allow air to provide lift and flight

aestivate to be dormant in summer (as compared to hibernating in winter)

Albian the last age of the Early Cretaceous epoch; 100 million to 113 million years ago

Aptian the age before the Albian; 113 million to 125 million years ago

alkaline having a low level of acidity

articulated jointed, as in a series of interconnected bones

bipedal using two feet

bivalve having two shells, such as a mussel

bryozoan belonging to a type of filter-feeding aquatic invertebrate animal also known as 'moss animals'

calcified hardened by, coated with or containing calcium carbonate

calcium carbonate a common chemical found in many rocks as well as living organisms (including eggshells, snail and sea shells, corals and teeth); the ability to use calcium to create hard structures was an important stage in the evolution of complex life forms

conifers trees and shrubs such as the pine tree that are usually evergreen and often produce seeds in cones

continental to do with a continent or large landmass

cosmopolitan found worldwide

crinoids a type of filter-feeding aquatic invertebrate animal such as the sea lily or feather-star, with arms radiating from around a mouth, and above a stalk

crustacean one of a group of aquatic animals including crabs, shrimps and lobsters

detritus the remains of living things

detrivores the creatures that feed on the remains of living things

dimorphic coming in two sizes, such as when one sex is larger than the other

dinoflagellates small, mostly marine, single-celled organisms such as plankton

dorsal of, or on, the back of an organism

electroreception the ability to sense electrical activity

endemic native to a particular place

escarpment a cliff-like ridge of rock

estuarine of a large often shallow coastal inlet; the type of organisms that live in it

etymology study of the origin of words

extinction event period of time characterised by the extinction of a very large proportion of living species

filamentous having hair-like structures, for warmth, protection or locomotion, etc.

filter feeder a species that sifts food from the water

fossils/fossilisation the traces of living organism often turned into stone (including bones, footprints, eggs, etc.); the formation of fossils

Gondwanan belonging to the supercontinent Gondwana which comprised Australia, Antarctica, Africa, India and South America

haemoglobin an iron-rich protein that transport oxygen in the blood and gives it its red colour

igneous rock formed through volcanic activity, as distinct from sedimentary rock (formed from the deposition of material, often in water) or metamorphic rock (which has been transformed through another process, such as heat or pressure)

invertebrates animals without a backbone

lineage a group of species that have evolved from one another

megafauna a diverse range of characteristically large animals such as mammoths and diprotodons that became extinct in the Pleistocene period

multicellular organisms that are made up of more than one cell (unlike bacteria or dinoflagellates)

nymphs the young of many insects (often an aquatic stage)

opalisation the process by which material such as a fossil is changed into opal

optic lobes part of the brain concerned with vision

oxygenated enriched with oxygen

palaeontologist someone who studies prehistoric life, such as fossils

sediments grains of mineral or organic matter deposited by water, air or ice

serrated having a toothed, or notched, edge

speciate the process by which a single species separates into a range of different species over time

stromatolites a species of colonial bacteria that creates mushroom-like structures in shallow salty seas (like Shark Bay, Western Australia); one of the oldest surviving forms of life

tectonic relating to the Earth's crust and the movements that cause earthquakes, folds and faults, giving rise to mountains and ocean trenches

terrestrial living on land

vertebrate an animal with a backbone

MUSEUMS

All Australian state museums have collections of marine fossil material from the Aptian–Albian as well as material from overseas and terrestrial material from the same period. Most also have displays of prehistoric life from the Mesozoic period. Museums with permanent displays of marine material from the Aptian–Albian include the following:

South Australian Museum, Adelaide — Opal Fossils Gallery — a permanent display of giant marine reptile and ammonite fossils. Features the Addyman Plesiosaur – a 6m long opalised skeleton found near Andamooka. http://www.samuseum.sa.gov.au/explore/museum-galleries/opal-fossils

Museums Victoria, Melbourne Museum — The Science and Life Gallery — features Dinosaur Walk with articulated skeletons of prehistoric reptiles, including pterosaurs flying overhead and displays of Mesozoic marine fossils and reconstructions. The IMAX theatre sometimes screens Sea Monsters 3D, which features the very similar Mesozoic marine life of America's inland sea. http://museumvictoria. com.au/melbournemuseum/whatson/current-exhibitions/dinosaur-walk/

Australian Museum, Sydney — Dinosaur Exhibition — features a wide range of species from the Mesozoic, including pteranodons and Eric the pliosaur, and a wide range of invertebrate aquatic fossils. http://australianmuseum.net.au/event/Dinosaurs

Australian Dinosaur Museum, Canberra features a range of marine reptiles from the Mesozoic including ichthyosaurs and pteranodons, as well as a wide range of fish, ammonites and nautiloids. http://www.nationaldinosaurmuseum.com.au/

Western Australian Museum, Perth — Diamonds to Dinosaurs exhibition? — features dinosaur skeleton casts and fossils of extinct life. http://museum.wa.gov.au/whats-on/long-term-exhibitions/diamonds-dinosaurs/#disqus_thread

Queensland Museum Southbank, Brisbane — Lost Creatures — features dinosaurs, megafauna and marine reptiles, including 3D modelling of Queensland dinosaurs, reconstructions, track ways of dinosaur footprints and marine reptile fossils. http://www.southbank.qm.qld.gov.au/Events+and+Exhibitions/Exhibitions/Permanent/Lost+Creatures

Many local museums also feature fossil material and displays from the Eromanga Sea.

Stonehouse Museum, Boulia — this historic house features an extensive display of local vertebrate and invertebrate fossils in the grounds. http://www.boulia.qld.gov.au/stonehouse-museum

Australian Age of Dinosaurs Museum of Natural History, Winton — this developing facility includes guided tours of the fossil collection, 3D animations, preparation opportunities and wildlife trails. http://australianageofdinosaurs.com/

Kronosaurus Korner, Richmond — the Fossil Centre features over 500 specimens and includes replicas of marine reptile species. Fossil preparation areas, audio-visual presentations, guided tours and children's activities are also available. http://www.kronosauruskorner.com.au/

Flinders Discovery Centre, Hughenden — this museum features a wide range of dinosaur fossils from around the world as well as local material, including a life-size replica of Muttaburrasaurus.

Further information on some of these museum can be found at the Dinosaur Trail. http://www.overlandersway.com/map/dinosaur_trail.aspx

ABC, 2014, The Eromanga Sea, The Age of Reptiles, <http://www.abc.net.au/science/ozfossil/ageofreptiles/eromanga/default.htm> viewed 16 November 2014

Australian Heritage Council, 2012, *Australia's fossil heritage, a catalogue of important Australian fossil sites*, CSIRO Publishing, Melbourne

Clode D, 2010, *Prehistoric giants: the megafauna of Australia*, Museums Victoria, Melbourne

Everhart M, 2009, Kronosaurus queenslandicus: *ancient monarch of the seas*, <http://www.oceansofkansas.com/kronosar.html> viewed 10 July 2012

Everhart, MJ, 2005, *Oceans of Kansas: a natural history of the Western Interior Sea*, Indiana University Press, Bloomington

Eyden, P, 2006, Tusoteuthis and the Cretaceous giant squids, *The Octopus News Magazine Online*. <http://www.tonmo.com/science/fossils/cretaceousGS.php> viewed 8 July 2013

Habib M, 2013, *Life in the air: pterosaur flight*, Pterosaur.net, Pterosaur flight, <www.pterosaur.net/flight.php> viewed 1 July 2013

Kaplan M, 2009, *'Sea monster' bones reveal ancient shark feeding frenzy*, National Geographic News, 28 October 2010, <http://news.nationalgeographic.com/news/2009/09/090928-fossil-shark-feeding-frenzy.html> viewed 4 July 2013

Kear BP and Hamilton-Bruce RJ, 2011, *Dinosaurs in Australia: Mesozoic life from the southern continent*, CSIRO Publishing, Melbourne

Long JA, 2011, *The Rise of Fishes: 500 Million Years of Evolution*, University of New South Wales Press, Sydney

Long, J, 1998, *Dinosaurs of Australia and New Zealand and other animals of the Mesozoic era*, Harvard University Press, Cambridge MA

Martyniuk MP, 2012, *A field guide to Mesozoic birds and other winged dinosaurs*, Pan Aves, Vernon NJ

McGowan C, 1991, *Dinosaurs, spitfires and sea dragons*, Harvard University Press, Cambridge MA

Stillwell J and Long J, 2011, *Frozen in time: prehistoric life in Antarctica*, CSIRO Publishing, Melbourne

Taylor PD and Lewis DN, 2006, *Fossil invertebrates*, Natural History Museum Publishing, London

Thulborn T and Turner S, 1993, An elasmosaur bitten by a pliosaur, *Marine Geology*, 18:489–501

Ward PD, 1987, *The Natural History of Nautilus*, Allen and Unwin, Sydney

Witton MP, 2013, *Pterosaurs: natural history, evolution, anatomy*, Princeton University Press, Princeton and Oxford

Witton, M, 2013, *From belly-dragging sprawlers to two-legged dynamos and everything in-between: terrestrial locomotion in pterosaurs*, Pterosaur.net, Terrestrial locomotion <www.pterosaur.net/terrestrial_locomotion.php> viewed 1 July 2013

Witton, M, 2013, *Food, sex and over-excess, pterosaur palaeoecology in a nutshell*, Pterosaur.net, Ecology <www.pterosaur.net/ecology.php> viewed 1 July 2013

AUTHOR'S BIOGRAPHY

Danielle Clode is a zoologist and an award-winning science writer. She has a doctorate in animal behaviour from Oxford University and has worked as a scientific interpreter, essayist and research fellow at Museums Victoria. Danielle is the author of several books, including *Killers in Eden* (2011) and *Prehistoric Giants: The Megafauna of Australia* (2009).

ACKNOWLEDGEMENTS

I would like to thank my publisher, Melanie Raymond, for co-ordinating, overseeing and supporting this project and many others. This book builds upon the expertise and efforts of countless scientists and artists whose careers are devoted to bringing these prehistoric creatures to life. We could not even begin to imagine this past world without all their work and I am indebted to their efforts. I am very grateful for the generous expert advice and assistance of Sam Arman, Alan Bartholomai, Bob Henderson, Scott Hucknull, Ben Kear, Mike Lee, John Long, Ben McHenry, Tom Rich and Maria Zammit, whose suggestions and contributions significantly improved the book, as did those of my editor, Nan McNab. My apologies for any remaining errors.

IMAGE CREDITS

Thanks are also due to the artists and scientists who have provided their images for this book, including Mary-Anne Binnie at the South Australian Museum. A special thanks to Lauren Nicholls, my own personal graphic artist and tutor. These images could never been compiled and used to such great effect without the considerable support of Museums Victoria staff: Melanie Raymond, Marija Bacic, Cathy Mulhall and Sally Rogers-Davidson. Thank you all.

The fishes on page 53 are drawn by the author and based on the following images originally published in various editions of the Queensland Museum Memoirs, © Queensland Museum. *Dugalida emmilta* (Lees, 1990), *Flindersichthys denmeadi* (Bartholomai, 2010), Cooyoo australis (Bartholomai & Lees, 1987), *Richmondichthys sweeti* (Bartholomai, 2010) and *Pachyrhizodus marathonensis* (Bartholomai, 2012).